YOUR KNOWLEDGE HAS VALUE

- We will publish your bachelor's and master's thesis, essays and papers

- Your own eBook and book - sold worldwide in all relevant shops

- Earn money with each sale

Upload your text at www.GRIN.com and publish for free

Energy Return On Investment with the concept of EROI. Applications, criticism and implications

Anna Szujo

Bibliographic information published by the German National Library:

The German National Library lists this publication in the National Bibliography; detailed bibliographic data are available on the Internet at http://dnb.dnb.de.

ISBN: 9783346382207
This book is also available as an ebook.

Abstract

Energy has a significant impact on economic growth and is a key driver for the wellbeing of a society. The less a society has to spend on energy, the more remains for consumption and discretionary spending that is directly translated into economic growth. This impact can be assessed with the help of the net energy analysis that makes use of the concept of EROI. The Energy Return On Investment is the ratio of the quantity of energy delivered to the quantity of energy consumed in a given process. Thus, this metric serves to measure the accessibility of a resource, meaning that the higher the EROI, the greater the amount of net energy delivered to society in order to support economic growth. This goes hand in hand with the finding that there is a minimum level of EROI that has to be reached, otherwise economic growth cannot be possible. Depending on the studies conducted, the minimum EROI for society is between 7:1 and 15:1.

Current research shows that the EROI for fossil fuels is on the decline, simultaneously leading to higher energy prices and thus impacting economic growth in a negative way. As a consequence, if the corresponding EROI is under the tolerable minimum, it is no longer profitable (both from an energetic and financial point of view) to extract that energy regardless of the reserves that are available. In order to find alternatives which are sustainable not only from an ecological but also from an economic perspective, EROI can be used to both comparing energy production technologies and deciding whether an investment in a specific technology contributes or not to the economic growth, in other words whether the technology produces energy with a level above the minimum EROI. The findings show that currently most of renewable energy sources do not reach the minimum EROI. Nevertheless, according to the concept of the ideal EROI there is considerable potential for improvement and thus the possibility to further increase the EROI.

Given that net energy analysis is going to be one of the most fundamental concepts in academic and policy discussion in view of the future of the energy mix there is still a clear need for a standardized and independent framework to calculate EROI.

Table of contents

Table of figures

1. Motivation

All forms of economic production and exchange require the transformation of materials, which in turn requires energy. The resulting role of energy and the significance of EROI relative to the global economy can be acknowledged both from a macroeconomic and microeconomic perspective.

1.1. Energy as an essential factor of production

Regarding the role of energy as a production factor, there is still debate among economists. Given that the importance of land as a production factor has been greatly reduced since the industrial development, today, the standard economic theory mainly distinguishes between two major production factors: capital and labor. After the two oil shocks in the 1970s and 1980s, energy was recognized occasionally as a distinct production factor by economists, however its importance in contributing to the economic production was generally ignored because of the low cost share of production (less than 5%). [1]

Georgescu-Roegen, Ayres and Kümmel were among the first and only ones who have under-stood energy as a production factor and called attention to the limits of growth with regards to the imminent scarcity of fossil fuels. In his work, Georgescu-Roegen [2] applied the second law of thermodynamics and, therefore, the concept of entropy for the analysis of economic systems, arguing that since all economic activities involve transformations of resources, they are inherently linked with inefficiencies and irreversible degradation [2].

In his model, Kümmel [1] also takes into account energy besides capital and labor. According to mainstream economics, output elasticity and cost share of the corresponding production factors have to be equal. Nevertheless, Kümmel shows that the output elasticity of a production factor, which measures the factor's productive power does not correspond to the respective cost shares of the total production costs. For energy these are much greater than the respective cost portions, while the opposite is true for labor. The productive power of energy is primarily given by technological constraints such as capacity utilization and degree of automation that are translated into monetary terms by shadow prices [1]. Thus it can be stated that energy accounts for a large part of the economic production which is ignored by mainstream economics and attributed to technological progress instead. Consequently, any changes into quantity or quality of energy have a significant impact on economic growth. One of the concepts to assess this impact is the EROI which serves as an indicator for energy prices and consequently, energy expenditure.

1.2. Energy surplus as a necessary criterion for survival

As emphasized a half century ago, every organism has to undertake activities that gain more energy than they actually cost in order to ensure their survival or evolution.

This "energy surplus" (or also called net energy) is given by the difference of the returned energy and "cost" for obtaining that energy and is considered a necessary criterion in order to allow the survival and growth of many species, including humans, as well as human endeavors, that is to say the development of culture, science and civilization itself. [3]

However, the question is not solely whether there is energy surplus, but also to what quantity, quality and what rate it is delivered which makes the difference from life-sustaining needs such as survival and maintenance towards additional functions being reproduction and evolution. Consequently, both biological systems and civilizations need to maintain a rather a substantial energy surplus than a bare one in order to ensure sustainable evolution. This is also true for contemporary industrial civilizations which are largely dependent on fossil fuels: These complex societies need a large quantity of energy resources with sufficiently high net energy (represented by a high EROI) in order to be able to be sustainable and to be wealthy from an economic, technological, cultural and educational perspective. As one can see, the EROI is not only an indicator for energy expenditure but also plays a major when it comes to determine whether a system is sustainable or not. [3]

2. EROI and its implications

2.1. The concept of EROI

2.1.1. Definition of EROI

Energy return on investment is the ratio of energy delivered to energy costs [4], in other words the ratio of how much energy is gained from an energy production process compared to how much of that energy (or its equivalent from some other source) is required to extract, grow, etc., a new unit of the energy in question [5]. EROI is strongly linked to net energy analysis which corresponds to the difference between energy input and output, and gives a statement on the energy surplus of a system. However, EROI is the more meaningful metric, because as a ratio it provides more information about the relationship of the input and output than a simple difference.

At the time of the introduction of the EROI concept, the core issue was rather whether the EROI of a technology was positive or negative and the main goal was to defend and defeat a particular technology instead of assessing and comparing various technologies in an objective manner. However, as stated above, for a system or a society to be able to sustain itself, it is important to have a minimum level of EROI and efforts have been made to introduce a standard framework in order to compare technologies in view of this condition.

The standard definition of EROI is given by [4]:

$$EROI = \frac{Energy\ gained}{Energy\ required\ to\ supply\ that\ energy}$$

The energy gained includes electricity, useable heat and power for useful work. The quantity of energy required to supply the energy gained includes drilling, refining, construction, installation, operations, maintenance, decommissioning, transportation, manufacturing of specialized equipment and chemicals required for extraction. The units are usually given in BTUs, kWhs or other units. As the nominator and the de-nominator are usually assessed in the same units, the EROI is dimensionless.

2.1.2. Consistent framework for EROI

One of the biggest issues with the EROI is that the methodology is not yet standardized which often leads to contradictory results. In order to achieve harmonization of EROI analysis a consistent framework has been introduced by Mulder and Hagens [6], which is given by two dimensions.

The first dimension corresponds to the depth of the analysis, i.e. whether and which factors need to be taken into account (direct and indirect inputs, externalities): The first order EROI, only dealing with direct inputs (energy and non-energy resources) and outputs, the second order EROI including indirect energy and non-energy inputs and also credits for coproducts (such as thermal content, mass, exergy) and the third order additionally gives account to the externalities linked to the technology which can be translated into additional costs and benefits. Externalities can be social, ecological and economic consequences for instance. Most commonly the lifecycle analysis (LCA) which is a method to calculate the impacts that are associated to the whole lifecycle of a technology is used to estimate the components of the EROI of a technology, especially second order EROI. The second dimension determines how to handle non-energy resources and externalities. Non-energy resources can include land, surface, ground water and time and often it is difficult for them to be translated into energy equivalents. Including these factors yields three possibilities: ignoring them, and hence resulting in the simple EROI, converting them into energy equivalents (called "total EROI") or dealing with them by defining separate indicators as additional components in the frameworks of a multi-criteria EROI. Given that energy is not going to be the only factor of production to be limited it is reasonable to focus on the third form of EROI. A higher order EROI means that the assessment becomes more comprehensive in scope and therefore more accurate, but there is a decrease in precision. [6]

2.1.3. Different versions of EROI

The narrowest definition is given by [4]

$$EROI = \frac{Energy\ gained}{Energy\ required\ to\ supply\ that\ energy}$$

Normally the EROI calculation is applied at the point at which the extraction or production plant is left. This corresponds to the concept of mine-mouth (also called well-head or farm-gate) and includes the energy to find and produce the fuel. This is the most commonly used form of the EROI assessment. However, it may ignore benefits or costs occurring after the point extraction.

The EROI can be calculated both for a specific technology but also for whole society in order to represent the importance of the EROI for an economic system: This version of EROI is called societal EROI and yields [3]:

$$EROI_{soc} = \frac{Summation\ of\ energy\ content\ of\ all\ fuels\ delivered}{Summation\ of\ all\ energy\ costs\ to\ get\ those\ fuels}$$

By applying the above mentioned framework by Mulder and Hagens the EROI analysis results in different versions of EROI: While $EROI_{mm}$ would correspond to the first order measure, the $EROI_{pou}$ which is taking into consideration the whole energy value chain including the energy to find, produce, refine and transport to the *point of use* corresponds to the second order definition. It is given by

$$EROI_{pou} = \frac{Energy\ returned\ to\ society}{Energy\ required\ to\ get\ and\ deliver\ that\ energy}$$

The most complete version of the EROI is the extended EROI which in addition to the other versions also include the energy used that are necessary for the infrastructure and transportation etc. [3]

$$EROI_{ext} = \frac{Energy\ returned\ to\ society}{Energy\ required\ to\ get\ and\ deliver\ and\ use\ that\ energy}$$

2.1.4. EROI trends for oil

In the 1930s during the oil development in Texas, Oklahoma and Louisiana it took on average one barrel of oil to find, extract and process 100 barrels of oil, which means that the EROI for in the US was about 100:1, then during the 1970s it decreased to 30:1, and at around 2000 it was between 11 to 18 returned per one invested. Thus, as we can see the EROI of our most important fuel is declining. The trend is similar for gas. Globally, the EROI was about 35:1 in the late 1990s and declined to about 20:1 in the beginning of the 2000s. With these trends continuing in about 20 to 30 years the EROI of oil and eventually gas will be equal to or lower than 1:1 which means that it will take at least one barrel of fuel to find and produce one barrel of fuel. In Europe the trends are similar. [7]

The following figure illustrates the trends of EROI for conventional and alternative energy systems [7]:

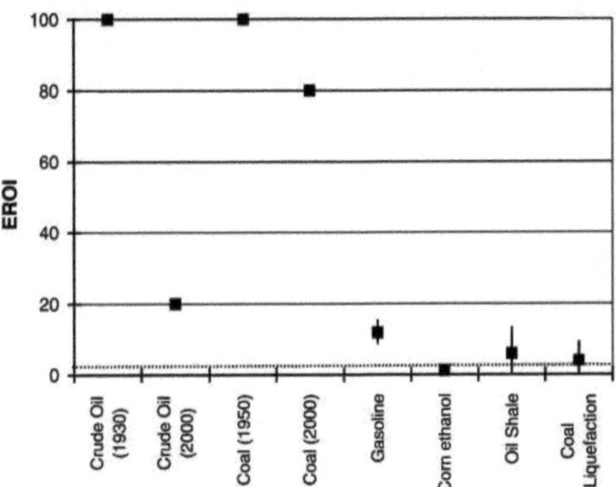

Figure 1 - EROI for conventional and alternative energy systems, Cleveland [7]

2.2. The minimum EROI for society

Due to the fact that the EROIs of energy production technologies is continually declining, it is crucial to determine to what point the EROI can decrease for a society to be still sustainable (minimum EROI). There are two approaches in order to estimate this level.

2.2.1. Bottom-up approach: Energy value chain

One approach to give an estimation for the minimum EROI from the bottom-up perspective has been proposed by Hall [3] by taking into consideration the value chain of energy production to determine the energy benefits and costs: For the case of oil as a fuel for a truck, $EROI_{mm}$ is currently estimated to 10:1 and the $EROI_{pou}$ is about 40 % less than the $EROI_{mm}$ (17% non fuel loss, plus 10% to run the refinery, plus 10% extraction, plus about 3% transport) indicating that at least for oil one needs an $EROI_{mm}$ of roughly 1.4 to get that energy to the point of final use [3].

By taking into consideration the energy services that are being delivered to the costumer ($EROI_{ext}$), the authors come to the conclusion that to provide the services of 1 unit of crude oil is roughly 3 units of crude oil is needed, similarly for other types of fuels. This means that the 10:1 $EROI_{mm}$ is cut to about 3:1 for a gallon at final use, since about two thirds of the energy extracted is necessary to provide the energy service at the end of the value chain. Thus a minimum EROI of 3:1 is required. In other words, to deliver one barrel of fuel to the final consumer and to use it requires about three barrels to be extracted. As a matter of fact, the EROI of 3:1 is only a bare minimum for civilization as it would only allow for the energy to reach the point of use but would leave only little discretionary surplus for things that use energy but do not contribute directly to getting more energy or other resources (art, medicine, education etc.). According to the authors, in order to maintain a civilization, it is estimated that an EROI would have to be considerably higher. [3]

The following figure illustrates the Society's pyramid of "Energetic Needs" representing the minimum EROI for conventional oil at the well-head (or mine-mouth) in order to perform various energetic tasks for civilization, analog to Maslow's Pyramid of Needs [8].

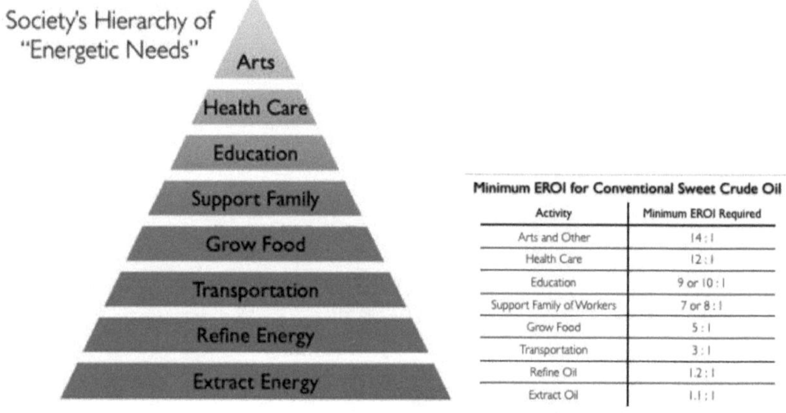

Minimum EROI for Conventional Sweet Crude Oil	
Activity	Minimum EROI Required
Arts and Other	14 : 1
Health Care	12 : 1
Education	9 or 10 : 1
Support Family of Workers	7 or 8 : 1
Grow Food	5 : 1
Transportation	3 : 1
Refine Oil	1.2 : 1
Extract Oil	1.1 : 1

Figure 2 - "Pyramid of Energetic Needs" representing the minimum EROI, Lambert et al. [8]

2.2.2. Top-down approach: Economic cost of energy

Works on the relationship of energy expenditure and economic growth have shown that in periods of higher energy costs, the expenditure is allocated from discretionary investments and consumption (mainly contributing to the GDP) to energy expenditure and that there is a threshold of energy expenditure above which the economic growth is no longer possible. Energy expenditure or the economic cost of energy is the "ratio of the cost of energy compared to the benefits of using it to generate wealth" which corresponds to the GDP [8]. To determine the above-mentioned threshold which is given by a percentage of the GDP, various studies have been conducted, one of the most recent and extensive one's by Fizaine and Court [9]. In their work they give an estimation of the maximum energy expenditure level above which economic growth would be impossible and transpose this result in order to find the maximum energy price and the minimum level of EROI that the energy sector must have in or-der to lead to a positive US economy's growth. With the help of a multi-variate linear regression model they come to the conclusion that for the US economy, only 11% of the GDP can be allocated to energy expenditure to still have positive growth which corresponds to a maximal energy price of 173 $2010 per barrel and a minimum EROI of 11:1. With regards to the energy price we are currently far from the threshold, however concerning the EROI the threshold is quite close. There are other approach-es as well, including the one of Weißbach et al. [11] proposing a minimum required EROI of 7:1 for the USA and Europe and the study by Lambert et al. [12], based on correlations between EROI and the Human Development Index (HDI), yielding a min-imum required societal EROI in the range 15:1 for contemporary societies.

2.3. Applications

2.3.1. Corn-based ethanol

One of the first major applications of EROI has been in the context of the corn-based eth-anol debate. The primary goal was to make a statement about whether ethanol based on corn has a positive energy yield (EROI greater than 1), and whether the break-even point is reached making the use of this technology worthwhile. If the energy returned is greater than the energy that is invested, then this is an indication that the investment should be made. However, as seen previously, there is a mini-mum EROI which has to be reached in order for an alternative to be worthwhile and not being subsidized by fossil fuels. The results of various analyses [5] show that the EROI of corn-based ethanol is between 1.2 and 1.6, although some works still show that the net energy return is not reached. EROIs above the bare technical threshold of 3:1 are rarely reported.

2.3.2. Comparison of alternative energy sources

Another application field for EROI is the comparison of different energy production technologies notably in view of the current energy transition and the intention to re-place the era of fossil fuels by renewable energy sources. Until this moment, major works have been conducted by Murphy and Hall [5]. Their results for fuels and alter-native energy sources are shown in the following figures. For further information, refer to Appendix 1.

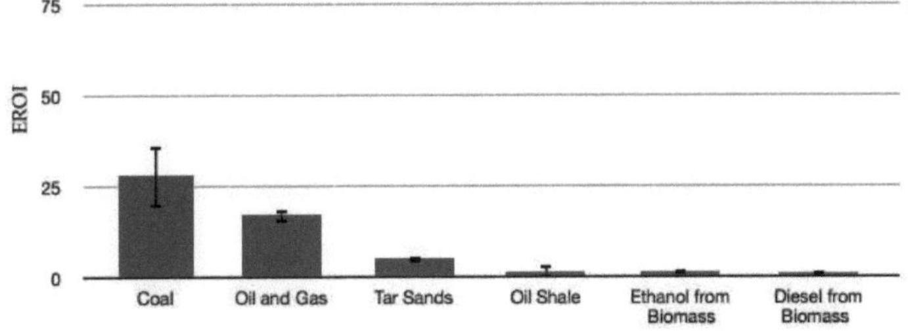

Figure 3 - Mean EROI (and standard error) values for thermal fuels based on known published, Hall et al. [10]

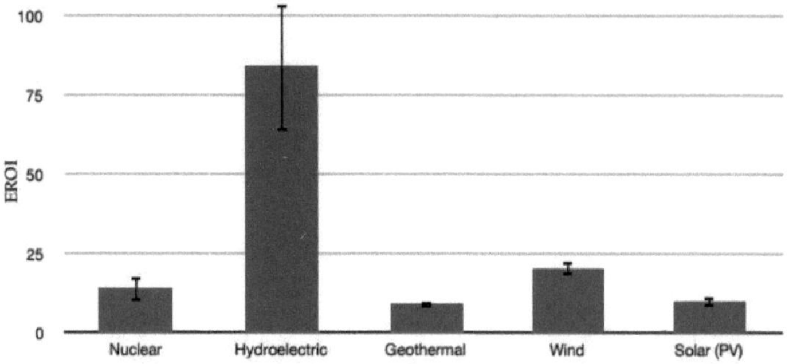

Figure 4 - Mean EROI (and standard error) values for known published assessments of power, Hall et al. [10]

In their corresponding work, Weißbach et al. [11] criticized the non-standardized EROI assessment and proposed another way that takes into consideration exergy in-stead of the primary energy in the EROI ratio. When considering renewable energy sources, besides the net energy analysis quality factors such as the usability have to be taken into account. In other words, all power systems have to provide exergy when it is required which means that the energy supply and demand have to be in accord-ance [11]. In order to compensate volatility effects of renewable energy sources, Weißbach includes the EROI of the neces-sary storage capacities to his assessment (buffering). In order to lower the EROI the least, the energy storage technology with the highest ESOI (Energy stored on energy invested) is used which is pumped hydro-electric storage. For other storage technologies, as it is the case for battery storage, the total EROI would be much lower. Weißbach's results show that nuclear, hydro, coal, and natural gas power systems have a significantly higher net energy surplus than photovoltaics and wind power and that most renewables do not reach the mini-mum EROI required by society. In Germany, for photovoltaic this limit is far from being reached. However, rooftop solar that is directly used and consequently not con-sidered as it is not powering society at large. [11]

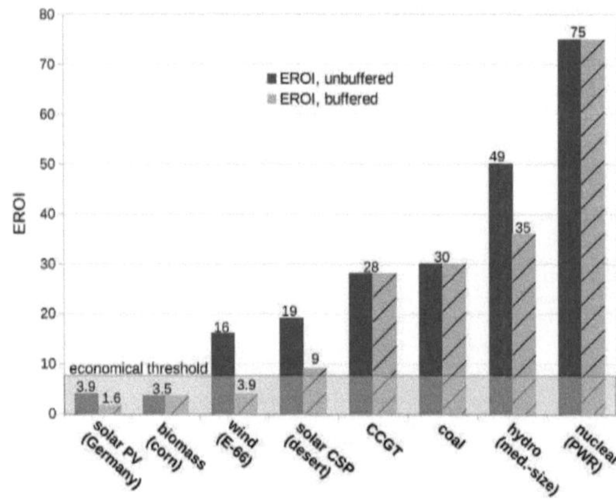

Figure 5 - EROIs of all energy techniques with economical "threshold", Weißberg et al. [11]

2.3.3. Ideal EROI

A new EROI metric has been proposed by Atlason et al. [13] named the ideal EROI, that provides the theoretical upper boundary of the EROI of a given system. It is given by the ratio between the energy inputs and the theoretical maximum output from a given system under the assumption that all losses are omitted and the lifetimes are longer. $EROI_{ideal}$ bears a resemblance to the concept of the idealized Carnot heat engine. Even though the EROI of a given system is never equal to the ideal EROI, this metric helps to estimate the potential for improvement and sheds a light on the efficiency im-provement possible for a given system. According to the calculations [13], the EROI of a wind turbine could double, hydroelectric power plants could improve approximately by 3 times, whereas the geothermal plant has a potential of 27-fold improvement over its lifetime (refer to Appendix 2).

2.4. Criticism

2.4.1. Lack of standardization

Although a standardized framework has been proposed by Mulder [6], this common protocol is often not being followed and there are still a lot of different EROI versions that are yielding strongly diverging results. One of the biggest issues regarding standardization is the boundary problem: it is often not clear which boundary and to which depth the analysis should go down the value chain and whether factors like human labor should be included or not. Often, there is still a discrepancy between the boundaries of the energy inputs and outputs. The resulting uncertainties are further emphasized by missing data (primarily due to missing funding sources).

2.4.2. Disregard of energy quality

One other flaw is that most net energy assessments do not take into account energy quality: normally the approach is to handle both inputs and outputs as thermal energy (Btu, joules etc.) although thermal energy equivalents not being equal since a Btu of electricity generates more economic output than a Btu of oil etc. The most common way to approach this problem is to apply the exergy analysis which uses the thermo-dynamics of energy conversion to assess quality of energy. However, this method is currently not always applied [4][14].

2.4.3. Lack of objectivity

Weißbach et al. [11] criticize that the results of the EROI of other works are often assessed to the detriment of conventional power sources, for example with regards to the lifetime: absurdly low lifetimes are assumed for fossil and nuclear plants, and un-realistic high ones for renewable plants. Another criticism point is that renewable energies have often been treated in a more favorable manner by weighting their out-put by a factor of 3 (motivated by the "primary energy weighting factor") but compar-ing it with the unweighted output of other energies like nuclear. Only a strict concept leads to independent and comparable results. Often the need for energy storage ("buffering") is not included thus resulting to higher EROI.

2.4.4. Insufficiency of the concept

Firstly, one critic point is that the EROI does not take into account the time factor in the calculations and hence the concept does not distinguish between a return from a short time interval or a long one. Other factors that are neglected by the EROI are greenhouse gases, energy independence and affordability. However, apart from the Energy Yield Ratio which has been proposed by Richards and Watt [12] for PV panels and thus offering a "better way of describing systems where different components may need to be replaced several times over the life of the whole system" [15], there is a lack of alternatives to the EROI.

2.5. Implications

2.5.1. EROI and economic growth

Throughout the history energy has been linked strongly to economic production: There is a statistically significant inverse relation between the US economic growth and the level of energy expenditure as a fraction of GDP between 1960 and 2010. As a consequence, increasing energy expenditure as a fraction of GDP can lead to a decline in US economic growth. [5]

Energy expenditure and energy prices in general are closely related to the EROI: A decline of EROI means that it gets harder and thus more expensive to produce energy and get it to the consumer. This can be understood like a feedback loop: If the EROI decreases due to the need to consume more oil in drilling deeper to oil resources, other inputs (e.g. steel) can become more expensive in both money and energy if they depend on oil for production and shipping. That means that as the price of oil gets higher, it can have a feedback making it more expensive to produce more oil. [5] As a consequence, decreasing EROI comes along with the increase of energy prices which will have a large impact on the economy given that a larger part of the GDP will have to be spent on energy and the people working in the energy industry instead of deliv-ering net dollars and energy to the rest of the soci-ety. This will hence impact discre-tionary spending and lower economic growth.

In 2015, the US GDP was about 12 trillion dollars, and it used about 100 quadrillion BTUs, which is equivalent to about 105 EJ [5]. Dividing the two we find that we use an average of about 8.7 Mega Joules to generate one dollar's worth of goods and ser-vices, 8,7MJ/dollar. The 105 EJ of industrial energy the US uses to run its economy for a year is equivalent to roughly 17 billion barrels of oil. At $70 per barrel that amount of oil would take 1.2 trillion dollars to purchase, which is about one tenth of GDP. This means that on average about 10 % of the dollar economy is used just to purchase the energy that allows the rest of the economy to function, which produces the end products we want. With current oil prices of $50 the energy expenditure would be at 0.85 trillion which amounts to 7% of the US economy. [5]

2.5.2. EROI and Monetary Return On Investment (MROI)

Research [16] also shows that besides the strong link between EROI and energy pric-es, the relation between EROI and the profit of energy-exploiting companies which is given by EMROI (Energy Monetary Return On Investment) is also considerable.

As it has been discussed above, there is an inverse relationship between EROI and the price of electricity such that as EROI decreases for a source of energy that energy becomes more difficult and thus expensive to obtain and the corresponding prices in-crease. From the point of view of a whole society, as the EROI falls an increasing pro-portion of the economy has to be devoted in order to obtain the same amount of net energy. As a conse-quence, there is less room for other investments.

On the other hand, EROI also limits the profit margin of a company: Over the long run, every energy-producing entity (EPE) has to make monetary profit in order to be able to survive in the long term (without subsidies). In other words, every EPE must sell its prod-uct energy for more than the monetary cost of the energy (direct and indi-rect) inputs required to produce it, this means that MROI > 1:1 and EROI > 1:1. [16]

The relationship between EROI and MROI is given by the following equation [16]:

$$MROI = \frac{EROI * p_i * e_{investment}}{e_i}$$

where e_i is the energy intensity (given by energy/physical quantity), $e_{investment}$ is the en-ergy intensity of investment (given in energy/money) and p_i is the unit price of produced energy. The relation this equation shows that the lower the EROI of energy systems, the less potential profitability for their businesses. As a consequence, the EROI can dictate the profit margin and the price necessary for a company to be involved in energy produc-tion which also means that EROI can be translated in a critical price range.

In their work, King and Hall [16] also come to the conclusion that energy production tech-nologies that are mostly dependent on their own product (e.g., fuels) are the cheapest. The implication of this concept are considerable for renewable energy tech-nologies that have very small consumption of fuel during their lifetime and are rather capital-intensive.

2.5.3. The paradox of oil

It is important to note that the above-mentioned conditions are equilibrium condi-tions with no constraints on any required inputs. In reality, there are changes in glob-al oil supply or demand that can have an influence on prices in a much faster time than the EROI itself. Thus, the concept of EROI can also be understood as giving a lower bound on price. This can be clearly seen with the current trend in oil prices where in spite of contin-ually declining EROI levels the price of oil is still decreasing because of the over-supply that is currently not met by the demand: On the demand-side, one can see that the global economy is deflated partly due to previously high oil prices and recession of economies dependent on fossil fuels and other commodities. While on the other hand, the supply has stayed on the same level and has led to large investments, overall resulting in a price drop. Low oil prices mean that the relative energy expenditure is also lower. With an average barrel price of $70 about 10 per-cent of the US economy is used for energy, with current oil prices of $50/barrel the energy expenditure would be at 0.85 trillion which amounts to 7% of the US economy [5]. As we have seen above lower energy expenditure leads to more options to invest in other parts of the economy in forms of discretionary spending which would in over-all have a beneficial effect on economic growth.

However, the drop in oil prices is affecting the profitability of energy producing com-panies given that the EROI is continuing to decline which translates into increased costs of production (capital expenditure has been rising 11% per year since 1999). Ac-cording to Alexander [17], approximately one-third of the current production is une-conomical with prices at $60, therefore having an impact on the economic stability of countries. He also states that due to increasing capital expenditure on new oil fields and declining EROI, it is nec-essary for oil to be sufficiently expensive so that supply can keep up with demand. But when oil gets too expensive, economies that dependent on cheap energy inputs cannot function and demand decreases, reducing the price of oil. This is also what happened all in all after the financial crisis of 2008 when oil plummeted and is currently on a very low level. Alexander argues that there is no longer a "narrow ledge" [18] where the price of oil is high enough to procure the nec-essary investments and production but not so high as to inhibit so-called 'healthy growth' of the economy and that oil is now either too cheap to procure ongoing in-vestments and production or too expensive for oil-dependent economies to function well (perhaps even both too cheap to meet demand and too expensive for growth). [17]

Nevertheless, in the last days OPEC countries and Russia agreed to cut oil produc-tion which means that there will certainly be a price increase coming and thus also affecting the profitability of oil producing entities.

3. Conclusion

As research shows EROI for fossil fuels is on the decline and will soon be below the minimum EROI for a society to sustain itself (ranging from 7:1 to 15:1) which according to the predictions could be reached before global fossil reserves are depleted. The implications of this are tremendous: With regards to the future availability of oil and other fossil resources, the question is not necessarily what the size of global oil re-serves is but rather what share of the reserves will be extractable with an EROI that ensures the sustainable development of society and what rate those fuels can be produced. The significance of these findings have to be borne in mind when seeking for alternative energy sources. As a matter of fact, with the current tendencies towards renewable power generation, the net energy analyses and the EROI becomes a very powerful metric for not only comparing different energy production technologies but also to make a statement whether technologies are worthwhile to invest in and thus representing an essential support for policy makers. In particular, because as we have seen the economic efficiency and wealth of a society strongly depends on the best choice of energy sources. As a consequence, the need to seek for possible substitutes that are in quantity and quality adapted to our economic system based on growth is urgent, especially against the background that renewable alternatives such as solar and wind they are below the minimum EROI. Thus the aim is to find an energy mix that in global in sustainable ecologically but also from the economic point of view being able to support modern society. At this point, the ideal EROI can be consulted to give an indication whether there is potential for improvement.

Energy is the backbone of economy. Energy prices and expenditure are directly linked to global economic production and economic growth and play a significant role in economic and financial crisis, potentially also the last one in 2008. For a high relative energy expenditure other parts of the economy will need to, less discretionary spending, less growth. EROI is the indicator to determine the relative cost of an energy production technology and is directly related to energy prices, meaning that declining EROI leads to higher prices. This is the reason why EROI may be the largest determinant of not only the ecological and energy-related aspects of our system but also our social, wellbeing etc.

Indeed, further investigation is going to be needed in order to assess and compare the EROI of different technologies in view of providing a sustainable economic system. However, it is to be noted that the EROI assessment should be independent from any political or ideological objective which seems to strongly influence the current debate. Nevertheless, the concept of EROI also has its limits as it is not the only metric having an impact on the economic production of a society, as we can see with the current oil price fall that is happening in spite of declining EROI. This means that other metrics also have to be consulted in order to be able to estimate the impact of an energy source in a holistic way.

It is also to be noted that the minimum EROI has been primarily calculated for developed countries, however developing countries the EROI is most likely on a lower level. So the question could be whether a lower EROI and thus particular energy sources would be sufficient and adapted for a developing economic system.

Bibliography

[1]   Kümmel, Reiner. "Economy." *The Second Law Of Economics*. 1st ed. New York: Springer Science+Business Media, LLC, 2011. Print.

[2]   Georgescu-Roegen, N. (1971). The Entropy Law and the Economic Process. The Economic Journal (Vol. 83). https://doi.org/10.2307/2231206

[3]   Hall, C. A. S., Balogh, S., & Murphy, D. J. R. (2009). What is the minimum EROI that a sustainable society must have? Energies, 2(1), 25–47. https://doi.org/10.3390/en20100025

[4]   Cleveland, C. J., & O'Connor, P. A. (2011). Energy return on investment (EROI) of oil shale. Sustainability, 3(11), 2307–2322. https://doi.org/10.3390/su3112307

[5]   Murphy, D. J., & Hall, C. A. S. (2010). Year in review-EROI or energy return on (energy) invested. Annals of the New York Academy of Sciences, 1185, 102–118. https://doi.org/10.1111/j.1749-6632.2009.05282.x

[6]   Mulder, K., & Hagens, N. J. (2008). Energy return on investment: toward a consistent framework. Ambio, 37(2), 74–79. https://doi.org/10.1579/0044-7447(2008)37[74:EROITA]2.0.CO;2

[7]   Cleveland, C. J. (2005). Net energy from the extraction of oil and gas in the United States. Energy, 30(5), 769–782. https://doi.org/10.1016/j.energy.2004.05.023

[8]   Lambert, J., Hall, C., Balogh, S., Poisson, A., & Gupta, A. (2012). EROI of Global Energy Resources Preliminary Status and Trends. College of Environmental Science and Forestry (NY), (November).

[9]   Fizaine, F., & Court, V. (2016). Energy expenditure, economic growth, and the minimum EROI of society. Energy Policy, 95, 172–186. https://doi.org/10.1016/j.enpol.2016.04.039

[10]   Hall, C. A. S., Lambert, J. G., & Balogh, S. B. (2014). EROI of different fuels and the implications for society. Energy Policy, 64, 141–152. https://doi.org/10.1016/j.enpol.2013.05.049

[11]   Weißbach, D., Ruprecht, G., Huke, A., Czerski, K., Gottlieb, S., & Hussein, A. (2013). Energy intensities, EROIs (energy returned on invested), and energy payback times of electricity generating power plants. Energy, 52, 210–221. https://doi.org/10.1016/j.energy.2013.01.029

[12]   Lambert, J. G., Hall, C. A. S., Balogh, S., Gupta, A., & Arnold, M. (2014). Energy, EROI and quality of life. Energy Policy, 64, 153–167. https://doi.org/10.1016/j.enpol.2013.07.001

[13]   Atlason, R., & Unnthorsson, R. (2014). Ideal EROI (energy return on investment) deepens the understanding of energy systems. Energy, 67(April), 241–245. https://doi.org/10.1016/j.energy.2014.01.096

[14]   Cleveland, C. J. (2005). Net energy from the extraction of oil and gas in the United States. Energy, 30(5), 769–782. https://doi.org/10.1016/j.energy.2004.05.023

[15]   Richards, B. S. Ã., & Watt, M. E. (2007). Permanently dispelling a myth of photovoltaics via the adoption of a new net energy indicator, 11, 162–172. https://doi.org/10.1016/j.rser.2004.09.015

[16]   King, C. W., & Hall, C. A. S. (2011). Relating financial and energy return on investment. Sustainability, 3(10), 1810–1832. https://doi.org/10.3390/su3101810

[17]   Alexander, S. (2015). The paradox of oil: The cheaper it is, the more it costs. Simplicity institute report 15a

[18]   Nelder, C., & Macdonald, G (2011). "There Will Be Oil, But At What Price?". Harvard Business Review. N.p., 2016. Web. 17 Dec. 2016.

Appendix 1

Table 1 - Published EROI values for various fuel sources and regions. Adapted from Lambert et al. [8]

Resource	Year	Country	Magnitude (EJ/ yr)	EROI (X:1)*	Reference
Fossil fuels (Oil and Gas)					
Oil and gas production	1999	Global	200	35	80
Oil and gas production	2006	Global	-	18	80
Oil and gas (Domestic)	1970	US	28	30	2, 24
Discoveries	1970	US		8	2, 24
Production	1970	US	10	20	2, 24
Oil and gas (Domestic)	2007	US	-	11	72
Oil and gas (Imported)	2007	US	28	12	6, 72
Oil and gas production	1970	Canada	-	65	83
Oil and gas production	2010	Canada	-	15	83
Oil, gas & tar sand production	2010	Canada	-	11	84
Oil and gas production	2008	Norway	-	40	81
Oil production	2008	Norway	-	21	81
Oil and gas production	2009	Mexico	-	45	82
Oil and gas production	2009	China	-	6	85
Fossil fuels (Other)					
Natural Gas	2005	US	30	10	152
Natural Gas	1993	Canada	-	38	83
Natural Gas	2000	Canada	-	26	83
Natural Gas	2009	Canada	-	20	83
Coal (mine-mouth)	1950	US	n/a	80	20, 48
Coal (mine-mouth)	2000	US	5	80	20, 48
Coal (mine-mouth)	2007	US	-	60	98
Coal (mine-mouth)	1995	China	-	18	198
Coal (mine-mouth)	2009	China	-	21	198
Other non-renewables					
Nuclear	n/a	US	9	5 to 15	20, 88
Renewables**					
Hydropower	n/a	n/a	-	>100	24
Wind turbine	n/a	n/a	-	18	77
Geothermal	n/a	n/a	-	n/a	70
Wave energy	n/a	n/a	-	n/a	70
Solar collectors**					
Flat plate	n/a	n/a	-	1.9	24
Concentrating collector	n/a	n/a	-	1.6	24
Photovoltaic	n/a	n/a	-	6 to 12	192
Passive solar	n/a	n/a	n/a	n/a	24
Biomass					
Ethanol (sugarcane)	n/a	n/a	-	0.8 to 10	151
Corn-based ethanol	n/a	US	<1	0.8 to 1.6	75, 75
Biodiesel	n/a	US	<1	1.3	73

* EROI values in excess of 5:1 are rounded to the nearest whole number.

** EROI values are assumed to vary based on geography and climate and are not attributed to a specific region/country.

Appendix 2

Table 2 - EROI results from various energy sources, Atlason et al.

Source/ Year	Hydro $EROI_{stnd}$	Hydro $EROI_{ide}$	Geothermal $EROI_{stnd}$	Geothermal $EROI_{ide}$	Wind $EROI_{stnd}$	Wind $EROI_{ide}$ Betz
1	3.3	4.1	17.0	52	1.4	2.8
5	15.1	20.3	28.2	230	6.9	12.8
10	27.6	40.2	30.7	400	12.7	23.4
15	38.1	59.7	31.7	530	17.5	32.4
20	47.0	78.8	32.2	633	21.77	40.1
25	54.7	97.5	32.5	717	25.4	46.8
30	61.4	116	32.7	786	28.5	52.6
35	67.3	134	32.8	844	31.3	57.8
40	72.5	151.7	33.0	894	33.8	62.3
45	77.1	169				
50	81.3	186.1				
55	85.0	202.8				
60	88.4	219.3				
65	91.5	235.4				
70	94.4	251.3				
75	97.0	266.8				
80	99.4	282.1				
85	101.6	297.1				
90	103.7	311.9				
95	105.6	326.5				
100	107.4	340.7				